ZERO POINT ENERGY PER STEREO RADIAN
AND THE DISTRIBUTION OF GRAVITATIONAL
ACCELERATION OF PLANETS THROUGHOUT

THE SOLAR SYSTEM

ZERO POINT ENERGY PER STEREO RADIAN
AND THE DISTRIBUTION OF GRAVITATIONAL
ACCELERATION OF PLANETS THROUGHOUT

THE SOLAR SYSTEM

The origin and cause of gravitation

Volume Two

Steven L. Basic

Library of Congress Control Number: 2013908608
ISBN: Hardcover   978-1-4836-3915-4
                 Softcover   978-1-4836-3914-7
                 eBook   978-1-4836-3916-1

This book was printed in the United States of America.

Rev. date: 07/22/2014

To order additional copies of this book, contact:
Xlibris
1-888-795-4274
www.Xlibris.com
Orders@Xlibris.com
599407

Contents

To my Daughter, Adrienna Faith Basic,

and

my Son, Christopher Michael Basic

Synopsis

Zero Point Energy (ZPE) is a subject of high interest in the scientific world. The Author has devoted time and efforts aimed at contribution in understanding and relationship of ZPE and the gravitational acceleration throughout the Solar System. The depth and power of his theoretical method directs some light onto previously unknown concepts. It is the results of Author's mathematical analytical achievement to demonstrate a circular frequency and ZPE's linear frequency and vortex-like simultaneous rotation. He named the particle ZPE-photon, and shows likely advance along a linear trajectory at the speed of light. The incorporeal ZPE waves are present in the entire Universe space at all temperatures. The fact is related to the left-hand side of Louis de Broglie equation.

The Author shows that the analytical magnitudes "g ZPE" gravitational fields and the observed "g" agree with a significant accuracy. The tables list valuable data, substantiation, also the locations of maximums and inflection points found at the distances between observed orbits and the center of the Sun. Therefore, the following conclusion becomes self evident: the intensities of the all surface gravities are dependent upon the angular velocities of all celestial spheres throughout the solar system. Previously studied "Occam razor" now is upgraded to a status of new Scientific Assertion.

Volume II Chapter 1

INTRODUCTION

Contemporary fundamental physics define four basic forces of nature as: (1) the gravitational force between mass particles, (2) the electromagnetic force between particles with charge including magnetism, (3) the color force between quarks, and (4) the week nuclear interactions that allow quarks to change their type related to neutron decays into a proton, an electron, and an antineutrino. The Author has devoted the intensity of focus to the gravitational fundamental physics forces that leads to understanding the quantum version of gravity and eventually defining the mass-less quanta of two units of spin called gravitons.

Previously the Author has published Volume One entitled *Unified Theory* (Reference 1). Following are Volume Two Chapters 1, 2, and 3. Specifically, the Volume Two defines a complete and exact mathematical algorithm indicating:

That the main Cause of Gravitation is the first main component, or, THE ZERO POINT ENERGY of Albert Einstein and Otto Stern, per 'Stereo Radian', the Zero Point Energy as it is, avoiding its more advanced form, because the Taylors polynomial of the first degree, approximates well all advanced forms. Furthermore, as the initial form had been reported in 'Analen Der Physik', 40:551, 1913, and as re-asserted by the most referable references of the Dean of the University of Leiden, Holland, Professor H. Casimir.

The EXPERIMENTAL discovery of professors Albert Einstein and Otto Stern of the Zero Point Energy equals to the simple product between the Max Plank constant 'h', and the frequency 'f', divided by the number two.

This phenomenon is a wave, therefore, at this stage, we suppose that this radiant source is located at the corona of the sun and illuminates the entire solar system in radial direction. This radiant energy will be shared with all the planets in proportion of planetary stereo radians.

AN EXACT MATHEMATICAL PROOF CONCERNING THE ZERO POINT ENERGY OF EINSTEIN—STERN PER STEREO RADIAN AND ITS EQUALITY TO THE STRENGTHS OF THE SURFACE GRAVITIES OF PLANETS THROUGHOUT THE SOLAR SYSTEM

CRUCIAL EXPERIMENT

(EXPEIMENTUM CRUCI)

Author: Steven L. Basic

In this monograph a previously unknown, very simple, relationship between the Zero Point Energy of A. Einstein—O. Stern and the Author's First Postulate (Reference No. 1), will be shown.

It is impossible to exclude the findings of French Scientist Louis De Broglie, A. Einstein and O. Stern from the High Technology, Science, Culture and Civilization. Therefore this relationship may be a crucial experiment—relating the known and previously unknown: the Zero Point Energy per stereo radian and the intensity of the surface gravitational fields, or, more precisely, the strain energies of the surface gravitational fields (Reference No. 2).

The first main postulate of the Reference No 1, reads:

$$E_{CONSTITUTION} = C^2 * \sqrt{\hbar * f}$$

Where:

$E_{CONSTITUTION}$: is the Energy of Constitution [kg*m]

C : is the velocity of light in outer space [m/sec]

$\hbar * f$: is elementary Radiant Energy, or, Zero Point Energy of A. Einstein and O. Stern, or, the right hand side of Louis De Broglie equation: $m * C^2 = \hbar * f$

[Kg*m, or, Joule]

$\hbar$: is the Max Planck constant [kg*m*sec, or, Joule*sec]

f : is the frequency [Hertz]

Since in all three cases the term: $\hbar * f$ is the radiant energy in nature, according to Reference No. 1, originating at the Corona of the Sun because the Corona has on its surface an elevated electrostatic potential of approximately two Million Volts, it is logical to examine the Absorption of this reactive Energy per a single stereo radian:

$$(\hbar * f) * (\frac{r^2}{R^2})$$

Where:

r : is the outer radius of a planet [m]

R : is the Orbital radius of a planet, or, its distance from the center of the Sun [m]

$(\frac{r^2}{R^2})$: is the Stereo Radian, by definition

The term under square root: $(\hbar * f) * (\frac{r^2}{R^2})$ will be denoted by "X", with this notation the first postulate will read:

$$E_{CONSTITUTION} = C^2 * \sqrt{X}$$

Taylor's (McLaurin) polynomial of the first degree, in respect to "X" as an independent variable, will read:

$$E_{CONSTITUTION} = C_1 + C_2 * X$$

Expressing the independent variable "X" with its original form:

$$E_{CONSTITUTION} = C_1 + C_2 * (\hbar * f) * (\frac{r^2}{R^2})$$

From the first principles of the theoretical mechanics and collected works of the Reference No.2, it follows:

$$\Omega = 2 * \pi * f$$

And according to the first principles of theoretical mechanics, a Total Resultant angular velocity of two staged relative rotation that's the first stage is Ω_1 and the second stage is Ω_2 is equal to simple SCALAR sum:

$$\Omega_{TOTAL} = \Omega_1 + \Omega_2 \quad \text{(SCALAR SUM)}$$

And since a relative rotations of Planets in respect to their polar axes Ω_1 , and also in respect to center of Sun Ω_2 are known and well established, the expression:

Zero Point Energy per Stereo Radian and the Distribution of
Gravitational Acceleration of Planets Throughout the Solar System

17

$$E_{CONSTITUTION} = C_1 + C_2 *^{*(\hbar * \Omega_{TOTAL})*(\frac{r^2}{R^2})}$$

becomes calculable.

To the greatest possible SURPRISE to the Author and the entire scientific world, the following causal relation is evident:

$$E_{CONSTITUTION} =$$
$$C_1 + C_2 *[Zero\_Point\_Energy\_of\_A.\_Einstein$$
$$and\_O.\_Stern\_per\_Stereoradian]$$
$$\approx It\_has\_been\_found \approx Suface\_gravities \approx g_{ZPE}$$

The relationship, with the substitution of the following, the first principles and Ref No. 2, from Page No. 4, equality:

$$\Omega = 2 * \pi * f$$

becomes:

$$E_{CONSTITUTION} \approx Surface\_gravities \approx g \approx$$

$$C_1 + C_2 *(\Omega_{TOTAL})*(\frac{r^2}{R^2})$$

With the following polynomial constants:

C1=8.136692
C2=1.250514*10^13

And the input data	Ω_{total} [RAD / SEC]	Planetary Radius [*10^6] m	Radius of Orbit [*10^11] m
Mercury	2.06527*10^–6	2.34	0.579
Venus	3.4299*10^–8	6.26	1.08
Earth	7.312*10^–5	6.37	1.49
Mars	7.098786*10^–5	3.32	2.28
Jupiter	1.773567*10^–4	69.8	7.78
Saturn	1.705567*10^–4	59.2	14.3
Uranus	1.613623*10^–4	23.7	28.7
Neptune	1.194612*10^–4	22.4	45.0
Pluto	1.138180*10^–5	3.0	59.1

ZERO POINT ENERGY PER STEREO RADIAN AND THE DISTRIBUTION OF
GRAVITATIONAL ACCELERATION OF PLANETS THROUGHOUT THE SOLAR SYSTEM

19

Resulted in the following 'g zpe' distribution of A. Einstein—O. Stern surface gravities of Planets throughout the Solar system:

	'g zpe' m/sec^2	g observed m/sec^2
Mercury	–	3.728
Venus	8.141	8.86
Earth	9.812	9.81
Mars	–	3.727
Jupiter	25.986	25.982
Saturn	11.78	11.369
Uranus	8.277	10.89
Neptune	8.177	11.87
Pluto	–	4.218

Since this calculation has been founded on the first

$$(m*C^2)_!$$

discovered by Professor A. Einstein, per stereo radian, or, the left hand side of Lois De Broglie equation, also, per stereo radian: $(m*C^2)_! *(\dfrac{r^2}{R^2})$ the magnitudes of 'g zpe' for planets:

Mercury, Mars and Pluto have been established with the second NON-RADIANT: $(m*C^2)_2$ that had been asserted by Madame Marie Curie at the end of nineteen Century:

$$(m*C^2)_2 \approx \frac{C_3}{r^2}$$

With the effect of this term 'g zpe' becomes:

$$g_{ZPE} = C_1 + C_2 * \Omega_{TOTAL} * (\frac{r^2}{R^2}) + \frac{C_3}{r^2}$$

In this case, the polynomial and integration constants are:

C1=10.09633
C2= 1.11303*10^13
C3=--7.20517*10^13

And the 'g zpe' and observed surface gravitation are:

	'g zpe' m/sec^2	g observed m/sec^2
Mercury	–	3.728
Venus	8.26	8.86
Earth	9.81	9.81
Mars	3.73	3.727
Jupiter	26.0	25.982
Saturn	13.3	11.369
Uranus	10.10	10.89
Neptune	9.99	11.87
Pluto	2.09	4.218

REFERENCES:

1.) *Unified Theory* by Steven Basic, ISBN 978-1-4349-9649-7, National Union Catalog OCLC. World Cat-Public View. Pending the Library of Congress Control number.

2). Lev Davidovich Landau, Collected Works.

3). *Indian Journal of Theoretical Physics*, Volume 47, No 1, ISSN 0019-5693, 1999, Institute of Theoretical Physics, Bignan Kutir, 4/1 Mohan Lane, Calcutta, 700 004, India, 'Some remarks concerning the distribution of gravitational acceleration and gravitational mass of planets throughout the solar system' by Steven L. Basic, Principal Engineer.

4). *'QUANTUM REALITY' BEYOND THE NEW PHYSICS AND THE MEANING OF REALITY, BY NICK HERBERT, ISBN 0-385-23569-0, ANCHOR BOOKS, DOUBLE DAY, NEW YORK,LONDON,TORONTO,SYDNEY,AUCKLAND*

5). ZERO POINT ENERGY PER STEREO RADIAN AND THE DISTRIBUTION OF GRAVITATIONAL ACCELERATION OF PLANETS THROUGHOUT SOLAR SYSTEM

Xlibris Corporation
Copyright 2013 by Steven L. Basic
Library of Congress Number   2013 3908608
ISBN   Hardcover   978-1-4836-3915-4
                        Softcover   978-1-4836-3914-7
                        Ebook   978-1-4836-3916-1
www.Xlibris.com
Telephone: 1–888–795–4274

6.) Lectures on Partial Differential Equations. Universite Pierre et Marie Curie (Paris 6). By Nicolas Lerner. February 24, 2011.

ZERO POINT ENERGY PER STEREO RADIAN AND THE DISTRIBUTION OF
GRAVITATIONAL ACCELERATION OF PLANETS THROUGHOUT THE SOLAR SYSTEM

23

ENERGY OF CONSTITUTION PER STERO RADIAN AND THE DISTRIBUTION OF THE PLANETARY SURFACEGRAVITATIONAL FIELDS THROUGHOUT THE SOLAR SYSTEM

By Steven L. Basic

Initially, the Author has asserted that the Zero Point Energy of A. Einstein and O. Stern per stereo radian generates the distribution that is identical to the observed distribution of the planetary gravitational fields (Reference No. 1and No. 2).

However, when developed into Taylor's polynomial of the first degree, the expression for the Energy of Constitution becomes proportional tothe Zero Point Energy of A. Einstein and O. Stern. Therefore, the Energy of Constitution per stereo radian should also be proportional to the observed distribution of the planetary gravitational fields.

An exact mathematical proof of this proportionality involves determination of the following term:

$$E_{CONSTITUTION} * (\frac{r^2}{R^2}) = \sqrt{C^4 * \hbar * f * (\frac{r^4}{R^4})}$$

Where, 'r': is the planetary shell outer radius;

'R': is the distance between a planetary orbit and the center of the Sun '(r/R) exp 2' : is the Stereo Radian; 'C' : is the velocity of light in outer space 'h' : is the constant of Max Planck; 'f' : is the frequency, $\Omega = 2*\varpi * f$

SUMMARY OF CONCUSIONS

The term under square root is determined for all nine planets and the Sun:

1.) The Energies of Constitution are proportional to the surface gravities of Planets and the Sun:

$$E_{CONSTITUTION} - per\_stereo\_radian \propto g$$

Where:
    'g' : is the planetary, or, star surface gravity.

2.) Commensurable nature between the Energy of Constitution and the Zero Point Energy.

REFERENCES

1. Basic, Steven L. *UNIFIED THEORY.* Lauriat Press. 2009. USA. ISBN: 978-1-4349-9649-7.

2. Basic, Steven L. Zero *Point Energy per Stereo Radian and the Distribution of Gravitational Acceleration of Planets Throughout THE SOLAR SYSTEM. The origin and causes of gravitation.* Xlibris LLC. 2013. USA. ISBN: 978-1-4836-3915-4.

3. Herbert, Nick. *QUANTUM REALITY. Beyond the New Physics . . . and the Meaning of Reality.* Anchor Books DOUBLEDAY. 1985. ISBN: 0-385-23569-0.

THE SOLAR SYSTEM IS A GRAVITATIONAL PLANETARY MASS—TIMES VELOCITY OF LIGHT SQUARED—SYSTEM

An Exact Mathematical Proof of Solar System
By Steven L. Basic

THE SOLAR SYSTEM IS A GRAVITATIONAL PLANETARY MASS—TIMES VELOCITY OF LIGHT SQUARED—SYSTEM

Background data. From the British Astronomical Association Handbook, the following most reliable data will be applied:

| 1.)

Planet	2.) Planetary outer shell radius 'r' meters x10 Exp6	3.) Observed surface gravitation 'g' m/sec2
Mercury	2.44	3.728
Venus	6.05	8.86
Earth	6.378	9.81
Mars	3.395	3.727
Jupiter	71.1	25.928
Saturn	59.65	11.369
Uranus	23.55	10.89
Neptune	24.2	11.87
Pluto	2.95	4.218

CELESTIAL SPHERES ZERO POINT ENERGIES (ZPE)

ZERO POINT ENERGIES OF ALL PLANETS THROUGHOUT THE SOLAR SYSTEM, DERIVED IN LIGHT OF REFERENCE No. 1., PAGE No. 65, FOURTH LINE FROM ABOVE, ARE GIVEN BY:

(ETERNAL WEIGHT 'W')*('r' EXP 2) = ZPE

IN THE CASE OF PLANETS, THE ETERNAL WEIGHT "W" FOLLOWS FROM THE GRAVITATIONAL MASS ("VIS GRAVITATIS"):

$$m_{gravitational} = \frac{r^2 * g}{\gamma}$$

AS THE PRODUCT:

$$m_{gravitational} * g = \frac{r^2 * g^2}{\gamma} = Eternal\_Weight = W$$

FURTHERMORE, FROM THE DEFINITION:

(ETERNAL WEIGHT 'W')*('r' EXP 2) = ZPE

THE ZERO POINT ENERGIES 'ZPE' OF PLANETS THROUGHOUT THE SOLAR SYSTEM WILL BE:

$$(ZPE)_{planets} = \frac{r^4 * g^2}{\gamma}$$

Where:

$'r'$: ARE OUTER PLANETARY RADII OF THEIR SHELLS

$'g'$: ARE THE SURFACE GRAVITATIONS OF THE PLANETS

$'\gamma'$: IS THE CONSTSANT OF THE UNIVERSAL GRAVITATION, FOR ALL THE PLANETS = 6.67EXP—11.

1.) Planet	4.) $(ZPE)_{planets} = \dfrac{r^4 * g^2}{\gamma}$	5.) $\sqrt{ZPE}$	6.) Zero Modulus= $\dfrac{m_{GRAVITATIONAL}}{\sqrt{ZPE}}$
Mercury	7.352Exp36	2.711Exp18	122194
Venus	1.569Exp39	3.961Exp19	122198
Earth	2.369Exp39	4.874Exp19	122201
Mars	2.756Exp37	5.249Exp18	122148

Jupiter	2.564Exp44	1.601Exp22	122191
Saturn	2.442Exp43	4.942Exp21	122179
Uranus	5.444Exp41	7.378Exp20	122170
Neptune	7.212Exp41	8.482Exp20	122140
Pluto	2.011Exp37	4.484Exp18	122182

FROM THE COLUMN No. 6, THE MAGNITUDES, IT IS POSSIBLE TO CONCLUDE THAT THE REST ENERGIES OF THE PLANETARY MASSES THROGHOUT THE SOLAR SYSTEM:

$$E_{REST} = m_{gravitatis} * C^2$$

ARE SIMPLY PROPORTIONAL TO THE FIRST POSTULATE OF FEFERENCE No. 1:

$$C^2 * \sqrt{ZPE}$$

WITH THE CALCULATED 'ZERO MODULUS':

$$m_{gravitatis} * C^2 = (Zero\_Modulus) * C^2 * \sqrt{ZPE}\_$$

OR,

$$m_{gravitatis} * C^2 = 122191 * C^2 * \sqrt{ZPE}$$

REFERENCES

1. Basic, Steven L. *UNIFIED THEORY*. Lauriat Press. 2009. USA. ISBN: 978-1-4349-9649-7.

2. Basic, Steven L. *Zero Point Energy per Stereo Radian and the Distribution of Gravitational Acceleration of Planets Throughout THE SOLAR SYSTEM. The origin and causes of gravitation.* Xlibris LLC. 2013. USA. ISBN: 978-1-4836-3915-4.

3. Herbert, Nick. *QUANTUM REALITY. Beyond the New Physics . . . and the Meaning of Reality.* Anchor Books DOUBLEDAY. 1985. ISBN: 0-385-23569-0.

SLB. June 16, 2014

Volume II Chapter 2

(2.1) A DESCRIPTION AND COMPONENTS OF THE ZERO POINT ENERGY VARIANCE

Introduction

At the beginning of the Twentieth century a prominent research Scientist and philosopher Nikola Tesla has determined that: '… a Rotation and a Natural Phenomenon of electricity are related as a Cause and Consequence…"

A parallel mirror image assertion is that: '…a Rotation and the surface gravities of celestial spheres throughout the entire solar system—planets—are related as a Cause and Consequence…" has been postulated one century later, by the author of this Second Volume, and the Unified Theory, Steve L Basic. A synthesis of the both assertions lead toward a conclusion that the surface gravities of celestial spheres—planets—are of electro-magnetic nature.

Since the Zero Point Energy of Albert Einstein and Otto Stern, is a WAVE, square root the first postulate of author, refer to this volume 2, is also a WAVE that has been found to be congruent to the planetary surface gravitation 'g'. By multiplying 'g' with any mass 'm' an expression for the Eternal weight will be obtained. When this weight is multiplied by the square of an 'extent':

('r*r), in light of the Reference 1, page number 65, the fourth line from above, then the variance at the left hand side agrees with the square of amplitude 'r' multiplier of any wave energy, per Reference No 3, page 75. Physical meaning of the left hand side as a product between the eternal weight 'W' and the square of an extent 'r*r' is the Zero Point Energy, whose previously unknown transcendental function at the right hand side has the following variables:

'W': is the weight [kg]
'r*r': is the square of the 'extent' [meters square]
'N': is the number of relative rotation stages [dimensionless]
'A': is the amplitude of the oscillating mass wave [meters]

In light of the second parallel assertion, the Second 'm*C square' has been found, that was hidden by nature, within the first, existing 'm*C square'.

The times of a single, complete revolution around the sun in years (Smithsonian Tables) of all the planets are simply proportional to an ascending Fibonacci sequence.

Since the Fibonacci sequence has golden proportions, the solar system is also of Golden Proportion. Reference 1 (page 22)

Please see Table 3 on the next page.

Table 3 [Please compare the magnitudes listed in column (12) and (10)]

(1)	(9)	(10)	(11)	(12)	(13)	(14)
Planet	$T_{OBSERVE}$ [sec *onds*] _or the _time_ _required_ for _one sin *gle* _complete revolution	$\dfrac{T_{OBSERVED}}{4,324,221}$ (4,324,221 -is_a_ common denominator_ of_ the_ column_(9)	^rA	Fibonacci Sequence $\dfrac{\phi^{r_A} - (-\frac{1}{\phi})^{r_A}}{\phi + \dfrac{1}{\phi}}$	(13) (12) * (8)	(14) (11) + (7)
Mercury	7,600,000	1.7575	3	2	3194	20
			4	3	2961	20
Venus	19,400,000	4.48635	5	5	3050	20
Earth	31,600,000	7.3078	6	8	3016	20
Mars	58,400,000	13.3737	7	13	3029	20
			8	21	3024	20
Asteroids			9	34	3026	20
			10	55	3025	20
Jupiter	374,000,000	86.4895	11	89	2937	20
			12	144	3024	20
Saturn	930,000000	215.489	13	233	3029	20
			14	377	3016	20
Uranus	2666,000,000	615.139	15	610	3015	20
Neptune	5200,000,000	1202.03	16	987	2961	20
Pluto	7820,000,000	1808.40	17	1597	3194	20

Where: $\phi=1.6180339...$is the Golden Proportion ('Sectio Auri')

Consequently, the product between the angular velocities of the mass points (planets) revolving around a central mass (the sun), and the logarithm "naturalis" of the angular velocities are a descending Fibonacci sequence (Reference No. 1, page No 20.)

Therefore, the product between both Fibonacci sequences is equal to the product of the energy and time (Professor Niels Bohr, Reference 28), this remains constant throughout the visible universe.

Reference No. 1 unifies the following two fundamental theories: The first theory is the historic Theory of Relativity of professor Albert Einstein, and the second theory is a contemporary Quantum Theory of professor Max Planck (Reference 8, and *Unified Theory* Equations 29 to 31, Page 18). This unification postulates a previously unknown relationship between the Energy of Constitution of Matter and the Frequency of any wave.

(2.2) THE MAIN (THE FIRST) POSTULATE:

$$E_{CONSTITUTION} = C^2 * \sqrt{h * f}$$

THE ENERGY OF CONSTITUTION 'E', BEING THE DIFFERENCE BETWEEN KINETIC AND POTENTIAL ENERGIES, EQUALS TO THE PRODUCT BETWEEN THE VELOCITIES OF LIGHT PROPAGATION IN OUTER SPACE— SQUARED, AND THE SQUARE ROOT OF THE PRODUCT BETWEEN THE MAX PLANCK

CONSTANT 'h' AND THE RESULTANT FREQUENCY OF A MULTI- WAVE ('f Hertz) (Definition 2.1).

According to the most referable references (Dean of the University of Leiden, Holland, professor H. Casimir)) the EXPERIMENTAL discovery of Professors Albert Einstein and Otto Stern of the Zero Point Energy that has been reported in'Analen Der Physik, 40:551, 1913', equals to the simple product between the Max Plank constant 'h', and the frequency 'f', divided by the number two. (Definition 2.2)

NOTE: The definition 2.2.) is located under the Square Root Symbol of the definition 2.1.).

Therefore, the EXPERIMENTAL definition 2.2. reflects a typical Taylor's Polynomial of the First Degree relative to the definition 2.1.:

$$E_{CONSTITUTION} = C^2 * \sqrt{2 * (\frac{\hbar * f}{2})}$$

Or:

$$E_{CONSTITUTION} = C^2 * \sqrt{2 * (E_{EISTEIN\_STERN}^{ZPE})}$$

(2.3) THE ZERO POINT ENEGY AND THE FREQUENCY DIFFERENCES

If the Zero Point Energy of Albert Einstein and Otto Stern, the State of No. 2 is denoted by ='h*f2/2' and the State No.1 ='h*f1/2', then the difference of the energy of Constitution 1.) reads,

$$(E_2 - E_1) = \Delta E_{CONSTITUTION} = C^2 * \sqrt{\hbar * (f_2 - f_1)}$$

Or

$$\Delta E_{CONSTITUTION} = C^2 * \sqrt{\hbar * \Delta f}$$

Hence

The "Frequency difference" of the energy of Constitution 'E', has been found to be congruent to the "Mass difference" of the rest mass, or, of the '(dm)*C^2'. This linkage served as a condition for a simultaneous development of the new causal relation with the Copenhagen interpretation of the "Quantum Theory."

(2.4) ASTROPHYSICAL SIGNIFCANCE OF THE HUBBLE'S DISCOVERY AND THE COPEHAGEN INTERPRETATION OF THE QUANTUM THEORY OF NIELS BOHR AND THE UNCERTAINTY PRINCIPLE OF WERNER VON HEISENBERG

When the Law of Hubble was discovered in twenties of the twentieth century by professor Hubble, the leading scientific advisor of the Pope and professor Albert Einstein have visited Hubble's observatory.

When the Uncertainty-like Principle of Werner Von Heisenberg for rotating mass-like events of Heavenly Spheres, (Ref. No 1, Page No 115, right hand side Quadrant, below):

$$(J * \Omega) * \frac{2 * \pi}{\Omega} = \hbar = Const\_1$$

is simply multiplied by the relationship of Hubble,

$$(\Omega)_{LEFT\_HAND\_SIDE-COFACTOR} = H_0 = (Const\_2)_{RIGT\_HAND\_SIDE\_COFACTOR}$$

It is interesting to note, Professor Hubble never claimed that the recession is radial exclusively, for example an observed spiral trajectory with nearly the same angle of the velocity vector, would be possible, then, the Copenhagen Interpretation of the Quantum Theory of Niels Bohr will be obtained:

$$J * \Omega^2 * \frac{2 * \pi}{\Omega} = \hbar * H_0 = Const\_3$$

If the first state of rotation is denoted as $(J_1 * \Omega_1)$; the second state of rotation $(J_2 * \Omega_2)$; and

$$T_2 = \frac{2 * \pi}{\Omega_2}, \quad T_1 = \frac{2 * \pi}{\Omega_1} \quad \text{then Niels Bohr relationship reads:}$$

$$(J_2 * \Omega_2{}^2 - J_1 * \Omega_1{}^2) * (\frac{2 * \pi}{\Omega_2} - \frac{2 * \pi}{\Omega_1}) = \hbar * H_0 = Const\_4$$

$$\Delta J * \Omega^2 * \Delta T = \hbar * H_0 = Const\_4$$

And the Uncertainty Principle of Werner Von Heisenberg for rotating mass-like events of Heavenly Spheres reads,

$$\Delta(J * \Omega) * \Delta T = \hbar = Const\_1$$

This result confirms an Astrophysical significance of Hubble, Niels Bohr and Werner Von Heisenberg relationships in light of an initial Werner Von Heisenberg—like analytical variance (Ref No 1, Page No 115, right hand side quadrant below.) (Professor Niels Bohr, Reference 28 of the Volume 1).

(2.5) THE FIRST EXPLANATORY NOTE

Recently published Transcriptional Monograph shows that THE ZERO POINT ENERGY of Albert Einstein and Otto Stern is in CAUSAL and EXACT relation with the energy famously asserted by Madam Marie Curie, reflecting the observed rotation of Celestial Spheres throughout the entire Solar system.

The fundamentals of THE ZERO POINT ENERGY (ZPE), are based on EXPERIMENTAL discovery of professor Albert Einstein and Otto Stern as reported in 'Analen Der Physik', 40:551, 1913. The Einstein-Stern findings state that THE ZERO POINT ENERY equals to the simple product between the Max Plank constant 'h', and the frequency of the ZPE wave 'f', divided by number two.

Many reputable sources confirm Einstein-Stern discovery, including the Dean of the University of Leiden, Holland, Professor H. Casimir, Wikipedia, and more. It is well known, the frequency of any wave 'f' according to the first principles equal 2*(3.14159...)*f = (Angular velocity 'Omega' of a corporeal, or, a Complex Rotation).

This fundamental relationship can be found, also, within principal works of Lev Davidovich Landau.

In continuation, by replacing the frequency 'f' with the Angular Velocity 'Omega' because of this fundamental proportionality, into the expression for the ZERO POINT ENERGY of Albert

Einstein—Otto Stern and further more and equating thus
obtained variance for 'ZPE' to the Energy of Madam Marie
Curie (that is: the second 'm*C square = equals = to a constant
divided by the square of an extent, or, the distance between two
points in space squared), constancy (invariance) of the AREAL
VELOCITY will be obtained.

The AREAL VELOCITY equals to a simple 'scalar' product
between the tangential velocity on its orbit (of a Celestial
Sphere) and the distance to the center of rotation (Sun center).

It is realistic to ask a question: IN NATURE are there Masses
that rotate in SPACE with a constant AREAL VELOCITIES
and why?

YES, there are: THESE MASSES ARE CELESTIAL
SPHERES (PLANETS) throughout the entire Solar System, or
The Second Law of Celestial Mechanics of Johannes Kepler.

PRIMER ON THE SUMMARY OF CONCLUSIONS

This EXPLANATORY NOTE confirms exactness of the 'ZPE'
of Professor Albert Einstein-Otto Stern, the Second 'm*C
squared' as asserted by Madam Marie Curie and the Second
Keplers Law of Celestial Mechanics.

Madam Marie Curie famously asserted that the energy may
depend, also, upon an 'extent' being the distance between two
points in space. It appears that this second 'm*C square' is
proportional to a constant divided by the square of the 'extent'.

According to the history of science and technology, the first, or, initial expression 'm*C square' has been discovered at the beginning of twentieth Century by Professor Albert Einstein. Many principal nations, the United States, Russia, China, France and others have built, with greatest possible efforts, their thermonuclear umbrellas, using the first, or initial 'm*C square' of professor Albert Einstein. Therefore it would be advisable to prove mathematically whether or not the second 'm*C square', famously asserted by Madam Marie Curie, DOES EXIST in NATURE.

On the first page, the right hand side a function under square root was found to be equal to the Zero Point Energy for the all planets throughout the entire solar system. It would be mathematically exact to replace the expression for the Zero Point Energy with a constant divided by square of a radial extent 'r', by taking square root of the energy of Marie Curie, the energy of constitution becomes proportional to a constant divided by the extent to the first degree, and because, according to theoretical mechanics, the first, derivative of the energy is equal to a Force—it is possible to understand this Force as Newtonian. Refer to the realm of the exploration of space. Both 'm*C square's' are exact—NOT MUTUALLY exclusive—nor in any way in contradiction.

This monograph is in support of the existence of a corporeal, parallel energy, that is additional to the first $(E = m * C^2)_1$ energy of Professor Albert Einstein, where 'E' is the energy; 'm' is the mass and 'C' is the velocity of light in outer Space.

ZERO POINT ENERGY PER STEREO RADIAN AND THE DISTRIBUTION OF
GRAVITATIONAL ACCELERATION OF PLANETS THROUGHOUT THE SOLAR SYSTEM

41

The second expression for '$(E = m * C^2)_2$' is inversely proportional to the square of an 'extent'. The 'extent' is the distance between two points in Space.

To quote Marie Curie who famously stated,

"EVEN WHEN NATURE OF SPACE BECOMES BETTER KNOWN, THE CAUSE OR REALITY WILL REMAIN A MYSTERY AND THE PHENOMENA DEPENDENT UPON AN EXTENT WILL, MOST LIKELY, ALWAYS PRESENT ITSELF TO US AS A PROFOUND AND WONDERFUL ENIGMA".

However, a transcriptional monograph Ref. No. 1, that had been published at the beginning of twenty first century, defines the Second, additional finding:

$(E = m * C^2)_2$ that does NOT exclude the first initial:

$(E = m * C^2)_1$ ONLY incorporates the energy that has been asserted, at the end of ninetieth century, by Marie Curie:

$$E_{MARIE-CURIE} = (m * C^2)_2 = -\frac{C^2 * C_3}{r^2}$$

where: 'r' is the 'extent', and C_3 is the constant of integration.

The best way of an exact proof of the existence of this corporeal, parallel energy, would be a mathematical comparison between the known

$$(E = m * C^2)_1$$

$$E_1 = m_1 * C^2$$

and

$$E_2 = m_2 * C^2 = -\frac{C^2 * C_3}{r^2} = E_{MARIE\_CURIE}$$

As their difference

$$(E_1 - E_2) = m_1 * C^2 - m_2 * C^2 = \Delta m * C^2 = RIGTHAND\_SIDE =$$

$$= m_1 * C^2 - (-\frac{C^2 * C_3}{r^2})$$

By taking $'m_1 * C^2'$ of the right hand side in front of the parentheses, the right hand side becomes:

$$m_1 * C^2 (1 + \frac{C_3}{m_1 * r^2})$$

However, surprisingly, in light of the Transcriptional Monograph, page No. 115, (Reference 1) for any value of a variable 'm':

Zero Point Energy per Stereo Radian and the Distribution of
Gravitational Acceleration of Planets Throughout the Solar System

43

$$\frac{C_3}{m * r^2} \equiv 1$$

Hence:

$$(\Delta m) * C^2 = 2 * m * C^2$$

The left hand side of this equation has been extensively explored with the experimental works since beginning of the Twentieth century, also, this is the case with the theory and the right hand side.

What was NOT known that the complimentary negative energy of Madam Marie Curie:

$$(-\frac{C^2 * C_3}{r^2})$$ is an integral component of corporeality:

$$(\Delta m) * C^2$$

(2.6) SOME REMARKS CONCERNING THE EXTENT OF AN ASSERTED 'NON-NUCLEAR' FISSION OF ELECTRONIC SHELL DURING THE RESONANCE AND ALTERNATING RADIATION

PRESSURE INTERRUPTIONS

The atomic radius of the electrolytic Aluminum is:

$$r_{ALUMINUOM}^{OUTER} = 0.149 * 10^{-9} \, meters$$

One half of this radius, or, midpoint of the electronic shell (cloud)

$$r_{ALUMINUOM}^{MEAN} = 7.155 * 10^{-11} \, meters$$

Refer to the 'Elements of materials science and engineering' Van Vlack, Addison Wesley Publishing Company, Page No. 54 (1980).

The second, additional, component of the rest energy

$$(E = m * C^2)_2$$

$$(E = \frac{C^2 * \hbar}{r^2})_2$$

At moderate velocities and far from the nucleus, in light of Reference 1, Page 115, where 'C' is the velocity of light in outer space;

$$C = 2.9979 * 10^8 \ \frac{m}{\text{sec}} \quad \text{and}$$

$$\hbar = 6.7539 * 10^{--35} \ Kg * m * \sec(in \ '\_ \ MKS' \_ system)$$

Refer to the 'The fundamental Atomic constants' by J.H. Sanders, Clarendon Press, Oxford, England, Page No. 79.

'r': is the location of an electron within the shell.

(2.7) UNIT CASE

In this particular example it would be reasonable to assume, that one single electron will be moved from the midpoint of Aluminum shell, or subjected to 'fission', towards an infinite distance 'r' and thus become a 'free' electron, in one second of time:

$$(E = \frac{C^2 * \hbar}{r^2} = \frac{8.987 * 10^{16} * 6.7539 * 10^{--35}}{(7.155 * 10^{-11})^2})_2$$

$$E_2 = 1185 \_ \frac{Kg*m}{\sec}$$

$CONVERTING\_TO'\_HP'(HORSEPOWER)$

$E_2 = 15.808\_HP$

$AND\_INTO\_KILOWATTS$

$E_2 = 11.539\_KILOWATT$

This amount of FREE power (11.539 Kilowatt) agrees with very many published experiments. However, this power may be somewhat greater or smaller depending upon the efficiency and the type of the interrupter design.

Since a moving charge (rapid jumps of electrons) across a spark gap that is subjected to a high voltage AC, generates an electromagnetic wave (which may or may not be heterodyning with other waves) causing a removal of electrons by radiation pressure, may well be simplest and perhaps the first, one hundred years known 'NON-nuclear' fission of Atomic electron shell.

(2.8) THE MAIN (THE FIRST) ASSERTION Per Unified Theory,

"SUN'S CORONA OF TWO MILLION VOLTS D.C., AMPLIFIES THE ZERO POINT ENERGY PHOTONS"

A proof of this assertion is reflected on page 62 of the Unified Theory, (Reference 1, page 62), on page 62, ABSORBED

ENERGIES BY ALL THE PLANETS OF THE SOLAR SYSTEM, and the Author's Table T2.

This table 'T2' is an 'Occam's razor', or, simple proportionality, without any constants. The multiplier 'A' that multiplies column '6' is equal=73,569,090. (dimensionless). In the column 8, observed surface accelerations of all the Planets are shown.

From the Table T2, next page) and the page No 93 of Ref No1, entitled 'The solar system of Albert Einstein and Otto Stern, it is possible to assert: 'THE SURFACE GRAVITIES OF PLANETS AND ROTATION ARE RELATED AS A CAUSE and CONSEQUENCE."—per this Volume 2 and the Reference No 1.

Table T 2

No	5	6	7	8
		$$E_C = C^2 * \sqrt{\dfrac{6.752398 Exp(-35)}{2*\pi}} * \dfrac{r}{R} * \sqrt{\Omega_{TOTAL}}$$	'6' *A This analysis 'Ec'	Observed
1	Mercury	1.71Exp-8		
2	Venus	3.17Exp-9		
3	Earth	1.15Exp-7	8.765	9.81
4	Mars	3.62Exp-8	2.813	3.727
5	Jupiter	3.52Exp-7	25.927	25.928
6	Saturn	1.57Exp-7	11.712	11.369

7	Uranus	3.09EXP-8		2.276	10.89
8	Neptune	1.80Exp-8			
9	Pluto	5.05Exp-10			

(2.9) SOME REMARKS CONCERNING THE DISTRIBUTION OF ETERNAL WEIGHTS OF CELESTIAL SPHERES THROUGHOUT THE ENTIRE SOLAR SYSTEM

In accordance with reference No. 1, and Author's monograph presented on July 7, 2011, at the Nikola Tesla Science Foundation, Philadelphia, the Mass Moment of Inertia Solution of an Euler-LaGrange differential equation, reads,

$$J = J_{NEWTON} * (\frac{C_1}{\Omega} + \frac{C_2 * \ln(\Omega)}{\Omega} + C_3 + \frac{C_4}{\Omega^2}) \quad \dots\dots\dots[1]$$

Where:

'J' : is the Mass Moment of Inertia

'J_{NEWTON}' : is the Newtonian Mass Moment of Inertia

$$= m_I * r^2$$

'm_I' : is the inertial mass ("Vis Insita").

'R' : is the orbital radius relative to a center of rotation.

'$\Omega = \dfrac{V}{r}$' : is the angular velocity.

'V' : is the tangential velocity of a revolving mass.

'$C_{1,} - C_2 - C_3 - C_4$' : are constants of integration.

Since: $J_{NEWTON} = m_I * r^2 = \dfrac{m_i * r^2}{\gamma} * \dfrac{\gamma * g}{g} = \dfrac{m_I * m_G * \gamma}{g}$ ....[2]

Where: '$m_G = \dfrac{r^2 * g}{\gamma}$' : is the Gravitational mass ("Vis
Gravitatis"].

'γ' : is the Newton's constant of Universal Gravitation.

'g' : are the surface gravities of Planets

Furthermore, Ref. 1: $m_I * m_g = \hbar * f$ .... [3]

Where:

'h' : is the constant of Max Planck
'f' is the frequency

From page No. 115, Ref. 1, and from Page No. 115.5 of this
Note:

J = Constant..[4]

Substituting equations [2], [3], and [4] into the equation [1], the equation No.1 will read,

$$J = \frac{\hbar * f * \gamma}{g} * (\frac{C_1}{\Omega} + \frac{C_2 * \ln(\Omega)}{\Omega} + C_3 + \frac{C_4}{\Omega^2}) \dots\dots\dots[5]$$

However, in light of reference No. 1 and No. 36 and the first principles:

$$\Omega = 2 * \pi * f \dots\dots\dots\dots\dots\dots\dots\dots\dots\dots\dots[6]$$

Substituting the relationship No. 6 into the equation [5]:

Equation No. [5] is valid for all values of 'J' including 'J=1' :

$$1 = \frac{\gamma}{2 * \pi * g} * (C_1 + C_2 * \ln(\Omega) + C_3 * \Omega + \frac{C_4}{\Omega}) \dots\dots[7]$$

Hence:

$$g = \frac{\gamma}{2 * \pi} * (C_1 + C_2 * \ln(\Omega) + C_3 * \Omega + \frac{C_4}{\Omega}) \dots\dots\dots[8]$$

It is very well known that in Nature a Constant velocity does exist. For example,

1.) 'C' a constant velocity of light propagation in Outer Space:

2.) Constant fraction of 'C' orbital of the electrons

3.) Simply 'Unity = 1'.

The Unity may involve assertion of a Constant tangential velocity Existence in Nature, and the equation becomes,

$$g = \frac{\gamma}{2 * \pi * D} * (C_3 + C_4 * \ln(r) + C_5 * r + \frac{C_6}{r})$$

'D=Denominator, that involves densities', from the Solution of an Euler-LaGrange differential equation (Ref. 2).

The second logarithmic term '$C_4 * \ln(r) * \frac{1}{D}$' can reverse its polarity, therefore will be denoted as Rutherford's term, to commemorate his early experimental "Pre-Fission" and "Pre-Fusion" studies in years about 1911.

The first, the third and the fourth analyzed (Ref.2) terms:

$$(C_3 + + C_5 * r + \frac{C_6}{r}) * \frac{1}{D} \quad \text{................................[10]}$$

have been in the Public Domain for two decades (Ref. 2). Their physical meaning has been found to be: the surface gravities of Planets throughout the entire Solar System: 'g'.

The work in reference 2 was peer-reviewed by the known Scientist Rahmanyan Chandrasekhar and his associates, and recommended for publishing, in the *Indian Journal for Theoretical Physics*, (1997 and 1999).

Therefore, the second term of Lord Ernest Rutherford,

$$C_4 * \ln(r) * \frac{1}{D}$$

was developed into the Taylor's polynomial of the first degree; its magnitude has been found to be proportional to the Surface Gravities of all the Planets through the entire Solar System. The symbol 'r' represents the outer Planetary radii.

(2.10). OUTER UNIVERSE

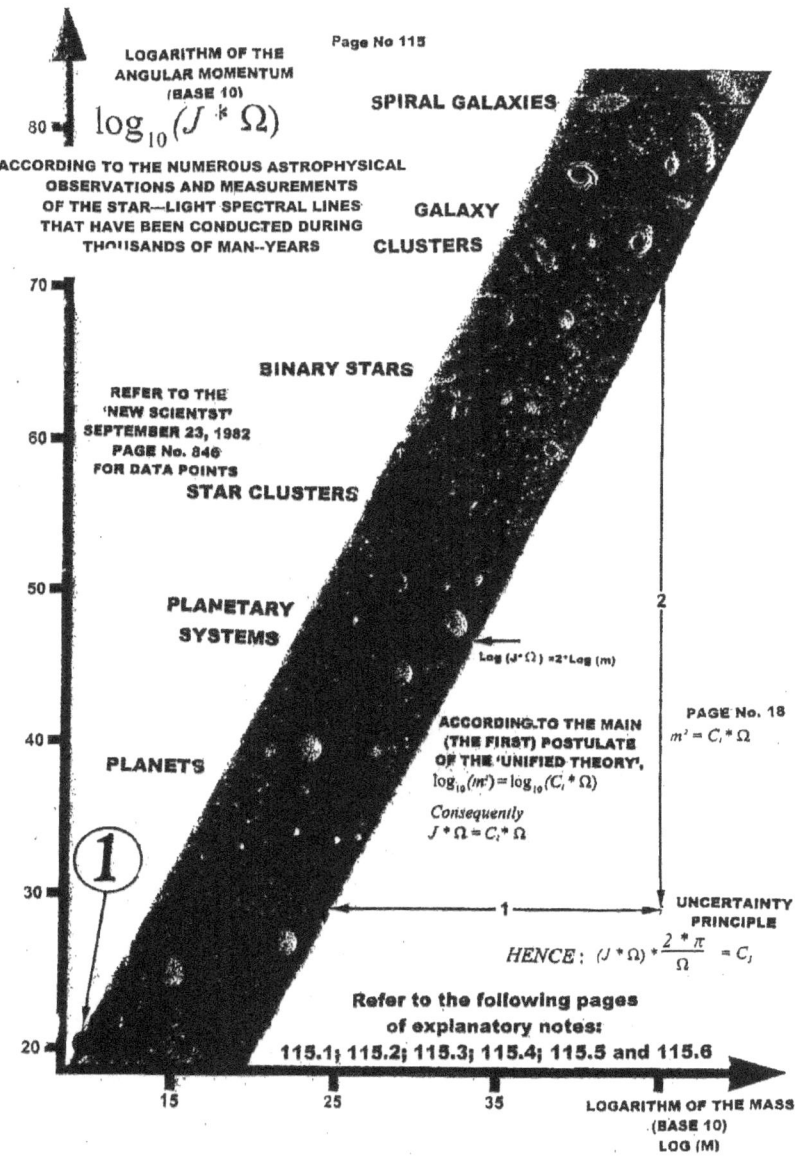

Page No. 115, from Reference No.1

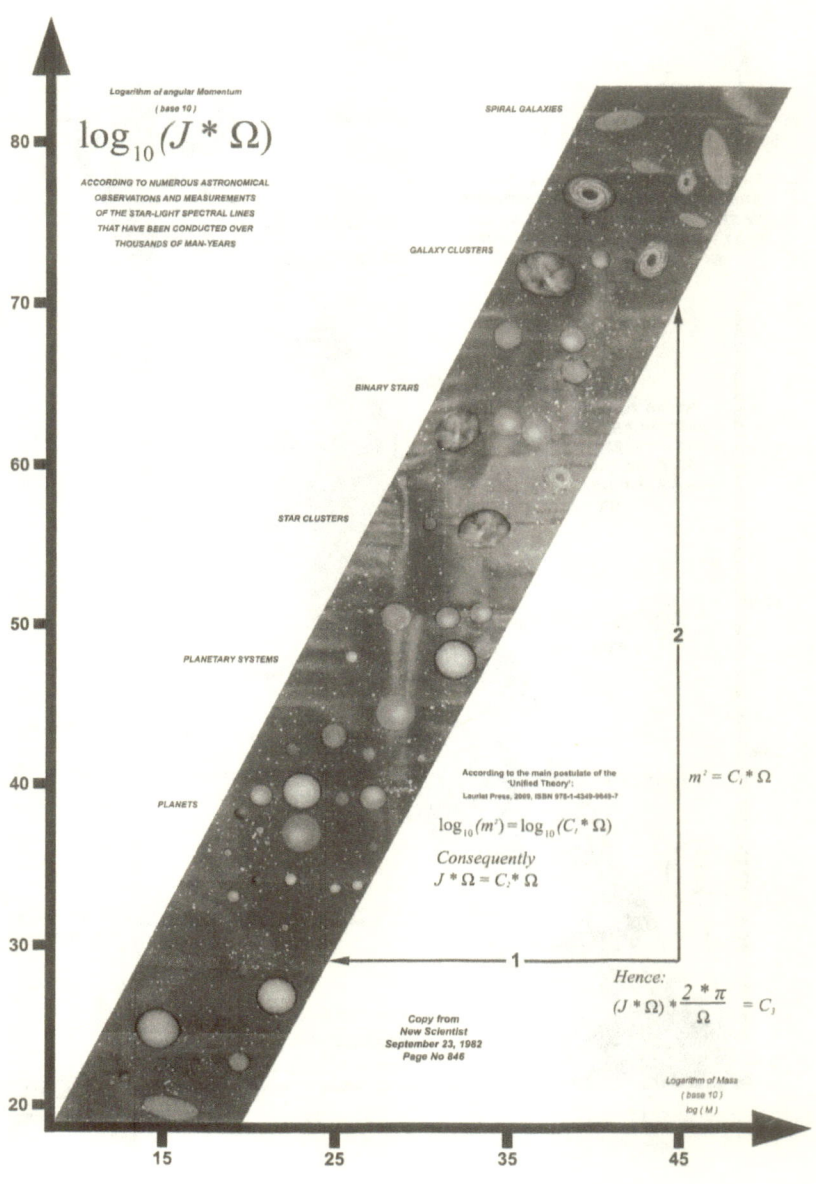

Logarithm of angular Momentum
(base 10)

$$\log_{10}(J * \Omega)$$

ACCORDING TO NUMEROUS ASTRONOMICAL
OBSERVATIONS AND MEASUREMENTS
OF THE STAR-LIGHT SPECTRAL LINES
THAT HAVE BEEN CONDUCTED OVER
THOUSANDS OF MAN-YEARS

SPIRAL GALAXIES

GALAXY CLUSTERS

BINARY STARS

STAR CLUSTERS

PLANETARY SYSTEMS

PLANETS

According to the main postulate of the
'Unified Theory':
Laurier Press, 2009, ISSN 978-1-4349-9849-7

$$\log_{10}(m^2) = \log_{10}(C_i * \Omega)$$

Consequently
$J * \Omega = C_i * \Omega$

2

$m^2 = C_i * \Omega$

1

Hence:
$(J * \Omega) * \dfrac{2 * \pi}{\Omega} = C_i$

Copy from
New Scientist
September 23, 1982
Page No 846

Logarithm of Mass
(base 10)
log (M)

80

70

60

50

40

30

20

15   25   35   45

(2.11) ASTROPHYSICAL SIGNIFICANCE OF WERNER VON HEISEBERG UNCERTAINTY—LIKE PRINCIPLE

With reference to the Diagram shown on the previous Page No. 115, the green point denoted by '1' (the left lower quadrant), has the following coordinates:

Abscissa 1: Log (m) =10, where 'm' is the mass and Log base 10.

Ordinate 1: Log= $(J^*\Omega)$ =20, where 'J' is the 'mass moment of inertia', 'Ω' is the angular velocity [radians per second] and Log based 10.

The second point 2 has coordinates:

Abscissa 2: Log (m) =0, m=1

Ordinate 2: Log $(J^*\Omega)$ =0, $J^*\Omega$ =1

An equation of a straight line passing through the points 1 and 2, also through the entire observed region, will read:

Log $(J^*\Omega)$ =2*Log (m) ....(1)

(2.12) A CAUSAL RELATION BETWEEN THE MAIN (THE FIRST) AND THE SECOND POSTULATES

Per Ref. No. 1, a photon like Nature of masses is shown

$$\frac{E}{C^2} = m_{PHOTON} = \sqrt{\hbar * f}$$

on the Page No. 18, Where:

'E': is the energy, 'C': is the velocity of light in outer space, 'h': is the constant of Max Planck and 'f' is the total frequency-angular velocity.

This mass will be substituted experimentally observed equation No. 1 (refer to Reference 1, Page No. 115)

$$Log\ (J^* \Omega\ ) = 2^*Log\ (m)\ \dots\dots\dots\dots\dots\dots\dots(1)$$

$$Log(J * \Omega) = Log(m^2) = Log[(\sqrt{\frac{\hbar}{2 * \pi}} * \Omega)^2]\ \dots\dots\dots(9)$$

$$Log(J * \Omega) = Log(\frac{\hbar}{2 * \pi} * \Omega)\ \dots\dots\dots\dots\dots\dots(10)$$

Since '$\frac{2 * \pi}{\Omega}$' term is the time for one single revolution around a central mass, the 'Uncertainty Principle' of Werner Von Heisenberg, for a rotating mass like event follows as the second postulate of the 'Unified Theory':

$$(J * \Omega) * \frac{2 * \pi}{\Omega} = \hbar \dots\dots\dots\dots\dots\dots\dots\dots\dots\dots\dots\dots\dots\dots(11)$$

Diagram No. 115 in reference 1, shown on the first, title page, and on the Pages No. 49 and 50 of this Volume No 2, reflects the work of a Supreme Intelligence. The data points were in public domain for over three decades but remained unacknowledged to a great extent.

> Please refer to the writing in *New Scientist*, page 846, September 23, 1982. A result of mathematical computations, related to an exact slope of this straight line (Ordinate = 2, Abscissa = 1, Reference 1, Page 115) is now available. All the symbols were added by Author's hand, demonstrate that the human mind CAN be in HARMONY with the INTENT of this Supreme Intelligence.

In the history of Science and Technology there is no record that the Surface Gravities of all the Planets throughout the Solar System are simply proportional to the Zero Point Energies of Albert Einstein and Otto Stern. However, within the Author's work an analytical proof of this proportionality is evident: the Taylors polynomial of the of the first degree of the Occam's razor is the Zero Point Energy of Albert Einstein and Otto Stern, refer to Authors monograph presented during the event in summer of 2011, in Philadelphia, at the Nikola Tesla Science Foundation.

WAVE ENERGY OF THE TWO STAGE OCILLATOR
Scientific Assertion by Steven L. Basic, Author

This scientific analysis is inspired on the text of Nick Herbert, the book *QUANTUM REALITY, Beyond the New Physics . . . and the Meaning of Reality.*

Professor Herbert's book has been greeted with high interest by academicians and scientist at the Universities of New York, London, Toronto, Sydney (AU), and Auckland (NZ).

On page No. 75, Reference 1, Professor Nick Herbert points to the significance of wave energy,

> *A WAVE'S EXTERNAL EFFECT DEPENDS ON THE ENERGY IT CARRIES, WHICH IS PROPORTIONAL TO THE WAVE'S INTENSITY (AMPLITUDE SQAURED).*

> *WAVE ENERGY GOES AS AMPLITUDE SQUARED. WHEN YOU DOUBLE A WAVE'S AMPLITUDE, YOU QUADRUPLE ITS ENERGY CONTENT.*

The importance of Herbert's statement is in the wave energy proportionality to the square of its amplitude.

This theorem is in complete harmony with the "ZERO POINT ENERGY" Ref. 3, and Professors Einstein and Stern throughout the Solar System:

$$(E_{ZPE}) * R^2 = (\frac{\hbar * f}{2}) * R^2 = (\frac{\hbar * \Omega}{4 * \pi}) * R^2 = (\frac{\hbar}{4 * \pi}) * (V * R)$$

"$\hbar$": is the Max Planck's constant

"f": is the frequency

"Ω": is the Planetary Angular Velocity in respect to the Center of Sun

"V": is the Planetary tangential velocity on its trajectory around the Sun

However, the assertion (V*R) = Constant, for each Planet, means: the Second Kepler's Law of Celestial Mechanics, or, that the "ZERO PONT ENERGY WAVE" of the planetary rotation does not change, the facts that point out to the Conservation of the Zero Point Energy.

Likewise, the Theorem of Nick Herbert is in complete agreement with the MULTI STAGE OSCILLATOR device designed at the Academy of Sciences (Milkovic), and with the Experimental Works at the Indian State University "Mahatma Gandhi".

At the University of India (Reference No. 4) the experimentally measured Total Power has been found to be:

28.05 Joule

1 Watt = 1 Joule/sec

In this Monograph an Ideal Total Power has been found to be: 25.16 Watts.

HERBERT'S POSTULATE APPLICATION TO THE GAIN, OR THE SURPLUS OF POWER OBTAINABLE FROM THE "TWO STAGE OSCILLATOR"

The Theorem of Nick Herbert is applicable to explain the performance of Two Stage Oscillator.

The Two Stage Oscillator (Milkovic) is presented in Diagram (DIa.) 1.

THE SECOND DRIVEN STAGE

"W 2"
The second driven Weight [Kg]

TWO STAGE OSCILLATOR--GEOMETRY

G
F
E
O
THE FIRST DRIVING STAGE
+Beta
A — · — D
C
B

"W1"
The first driving Weight [Kg]

Dia. 1)

The Amplitude of the First Driving Stage is the distance between the points "B" and "D" (Refer. to Dia. 1), or, one quarter of the full oscillation = (BD)/2.

However, the Amplitude of The Second Driven Stage is equal tp distance between poiny "F" and "G", or, equal (FG)/2

By sliding the The First Amplitude correspomding half of an wave "BD" so, that the point "D" coincides with the point "O", it is possible to establish a relationship between the total displacement of the First Driving Stage "BD" and the total displacement of the Second Driven Stage "FG": because of similarity of triangles "FOE" and "EDB", since the angles at "E" are equal at the both sides bar "FOD", the displacement "FG" follows:

"FG"=(FE)*(BD)/(OD)

$$\text{Since } \frac{A2}{A1} = \frac{\dfrac{FD}{2}}{\dfrac{BD}{2}} = \frac{FD}{BD}$$

With reference to Professor Herbert, Work done for one quarter of a Sinusoidal oscillation at the projection of the First Stage Driving Wave on the line "OBD", is equals to the displacement of the weight "G1" from the point "D" to the point "B", or, the Mechanical work done ="DB"*G1

Wave Energy of the First stage, according to the professor Herbert will be:

$$E_{WAVE-ENERGY\_1} = (DB * G1\ ) * (DB)^2 = G1 * (DB)^3$$

THE SECOND DRIVEN STAGE

A real movement lf the Second Driven Stage for one quarter of the full wave of the First Driving Stage is equal:

$$(FG) = (BD) * \frac{(EF)}{(OE)} [Metres]$$

[Relationship $\frac{(EF)}{(OE)}$ could be between "1" and "2"].

From this Wave energy of the Second Driven Sage, follows:

$$E_{WAVE+ENERGY\_2} = (G2 * FG) * (FG)^2 = G2 * (FG)^3$$

Total Wave energy of both Waves is equal to the SKALAR SUMM:

$$E_{TOTAL\_WAVE\_ENERGY} = E_{WAVE\_ENERGY\_1} + E_{WAVE\_ENERGY\_2} =$$
$$G1 * (BD)^3 + G2 * (FG)^3$$

The basic Mechanical Energy has been defined in Kilogram * Meters; that will be converted into Watts by multiplying basic Mechanical Energies with (1000/75/1.37).

According to the Research and Development results (Milkovic), the frequency of free oscillations of the First Stage is equal to f1 and the Second Stage to f2.

An Energy multiplied by the frequency defines the Power of the System: a higher frequency means a larger number of the Wave Energy Cycles.

$$P_{TOTAL} = 9.732*[G1*(BD)^3*f1 + G2*(FG)^3*f2]\{ Vat\}\_ili$$

$$= 9.732*G1*f1*(BD)^3*[1+(\frac{G2}{G1})*(\frac{f2}{f1})*(\frac{FG}{BD})^3]\_Vat$$

EXAMPLE

Reference No.1:

Asian Journal of Science and Technology, No. 4, Issue 08, pp 037—041, August 2013

Reference No. 2:

Research and Development Center of Academician V. Milkovic.

1. (DB) = 0.164 Meters

2. Weight G1 = 16 Kilograms of force (Reference No. 1)

3. Ratio between Driven Weight G2 and Driving Weight G1, ratio, or, G2/G1=1.5 (Reference No. 2)

4. Ratio between the frequency of the Driven Stage f2 and Diving frequency f1, f2/f1=2

(Reference No. 2)

5. Ratio of the displacement "FG" the Second Driven Stage and "BD" of the Driving Stage: (FG)/(BD)=2 (Reference: this monograph, Page No. 4)

6. $\dfrac{(EF)}{(OE)} = 2$

7. Frequency = 0.733 Hz (Reference 1)

8. There are two displacements "DB" per one single full oscillation

(and several oscillations before a new drive power is required)

With this set of independent variables, it follows:

Redoubling "2" was applied, since the passage of the weight W1 along "BD" happens twice per one single oscillation

$$P_{TOTAL} = 2*9.732*16*0.733*(0.164)^3*(1+1.5*2*(2)^3 =$$
$$= 25.16\_Watts / Per\_One\_Complete\_Oscillation$$

The ratio between the driven and driving Power, or, "Over Unity" ratio, in this example is twenty four.

It may be of some interest to note, that in this monograph only the Peer Reviewed Theorem of professor Nicholas Herbert was applied.

[1 Watt=1 Joul/sec]

Also, it would be necessary to note that the agreement with the measured increase in power at the University "Mahatma Gandhi", with the analytical results presented in this monograph, is excellent.

Author: Steven L. Basic

May 30, 2014

REFERENCES

1. Herbert, Nick. *QUANTUM REALITY. Beyond the New Physics . . . and the Meaning of Reality.* Anchor Books DOUBLEDAY. 1985. ISBN: 0-385-23569-0

2. Basic, Steven L. UNIFIED THEORY. Lauriat Press. 2009. USA. ISBN: 978-1-4349-9649-7

3. Basic, Steven L. *Zero Point Energy per Stereo Radian and the Distribution of Gravitational Acceleration of Planets Throughout THE SOLAR SYSTEM. The origin and causes of gravitation.* Xlibris LLC. 2013. USA. ISBN: 978-1-4836-3915-4.

4. *Asian Journal of Science and Technology,* No.4, Issue 08. Pp 037—041, August 2013

5. Research and Development Center of Veljko Milkovic (available upon request)

A FUEL—LESS—MULTI—STAGE—OSCILLATOR
MECHANICAL ENERGY AMPLIFIER OF
ACADEMECIAN MILKOVIC

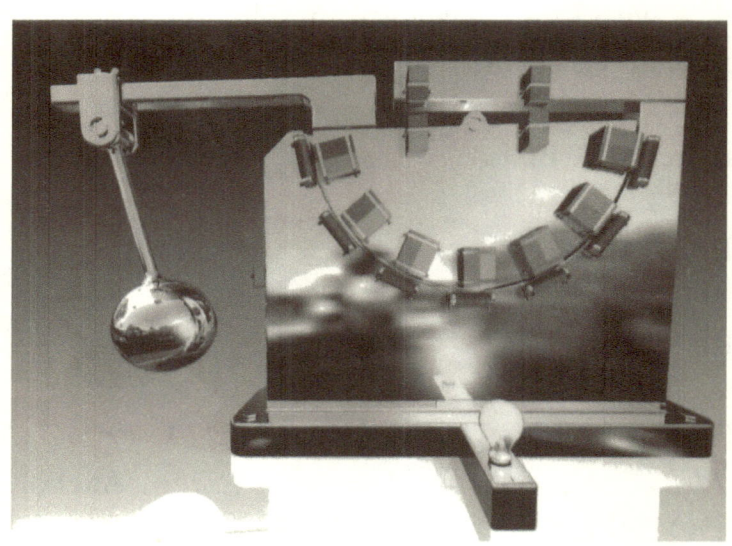

Volume II Chapter 3

CHAPTER 3 OBJECTIVE

The main objective of the Chapter Three is to provide an exact mathematical proof that the square of the inertial masses of all the celestial spheres (planets) is equal to the ZERO POINT ENERGY of Albert Einstein and Otto Stern.

The ZERO POINT ENERGY definition as defined in Unified Theory (Ref. 1), Page 65

$$E_{zpe} = W * r^2 \equiv m_i * \frac{g * r^2}{\gamma} * \gamma \equiv m_{INERTIAL} * m_{GRAVITATIS}$$

$$SINCE \_ m_{INERTIAL} \_ IS \_ PROPORTIANAL \_ TO \_ m_{GRAVITATIS}$$

$$\therefore \_ E_{ZPE} = m^2 \therefore m = \sqrt{E_{ZPE}} \therefore m * C^2 = E \_ = C^2 * \sqrt{E_{zpe}}$$

$$\_OR \_ THE \_ FIRST \_ POSTULATE \_ OF \_ REF. \_ NO. \_ 1$$

$$SINCE \_ : \frac{r^2}{R^2} \_ IS \_ THE \_ RATIO \_ BETWEEN \_ SQUARES$$

$$OF \_ THE \_ PLANETARY \_ RADIUS' \_ r' \_ AND \_ THE$$

$$RADIUS \_ TO \_ CENER \_ OF \_ THE \_ SUN' \_ R' \_ OR, \_ EQUAL$$

$$TO \_ THE \_ PLANETARY \_ STEREO \_ RADIAN, \_ AND$$

$$2 * \pi * f_{REQUENCY} = \Omega_{ANGULAR\_VELOCITY}$$

The following expression for the Energy of constitution will be obtained:

$$E = C^2 * \sqrt{\frac{6.752398 Exp(-35)}{2*\pi}} * \frac{r}{R} * \sqrt{\Omega_{TOTAL}}$$

The first term of this expression may be calculated:

$$The\_first\_term = C^2 * \sqrt{\frac{6.752398 Exp(-35)}{2*\pi}} = 0.294627682$$

Other terms for the celestial spheres (Planets) of the entire solar system are:

(Table T1, the same as on page 10 is repeated for easier reference)

Astrophysical Data

Table T1

No	1	2	3	4
	Planet	Ω Omega (total) [rad/second]	Planetary Radius In millions of Meters (*10^6)	Radius of the Orbit around the Sun Meters (*10^11)
1	Mercury	2.06527 Exp $^{-6}$	2.34	0.579
2	Venus	3.44299 Exp $^{-8}$	6.26	1.08

3	Earth	$7.312013 \, Exp^{-5}$	6.37	1.49
4	Mars	$7.098786 \, Exp^{-5}$	3.32	2.28
5	Jupiter	$1.773567 \, Exp^{-4}$	69.8	7.78
6	Saturn	$1.705567 \, Exp^{-4}$	59.2	14.3
7	Uranus	$1.613623 \, Exp^{-4}$	23.7	28.7
8	Neptune	$1.194612 \, Exp^{-4}$	22.4	45
9	Pluto	$1.138180 \, Exp^{-5}$	3	59.1

Planets: Circular frequency of the celestial spheres.

(by definition and according to the first principles)

$$\Omega = 2 * \pi * f \, -$$

-

$'f'$: is the frequency [Hertz]

Omega total 'Ω' = scalar, simple, and the sum of the following two, the main angular velocities:

C.) In respect to the center of the sun
D.) In respect to the planetary polar axis

The dimension of the Omega total is: [Radians per second]

(2.13) RADIANT ENERGIES ABSORBED BY ALL THE PLANETS OF THE SOLAR SYSTEM

The Author finds considerable importance in the radiant energies absorbed by planets. The Solar System may be observed through the frame set by Albert Einstein and Otto Stern, and in light of Leonard Euler's and Werner von Heisenberg radial acceleration of the random Zero Point Energy Photons generated by two million Volts potential at the surface of the Sun's corona. In his study the Author cannot disregard the fact that, the radiant energies are absorbed by all the planes in the Solar system.

On the next page is the Author's Data Table T2, where,
Column 5 represents planet name
Column 6 represents the values obtained by equation on column top
Column 7 represents the multiplier "A" = 62,801,558 (dimensionless)

Table T 2

No	5	6	7
		$$E = C^2 * \sqrt{\dfrac{6.752398 Exp(-35)}{2*\pi}} * \dfrac{r}{R} * \sqrt{\Omega_{TOTAL}}$$	(Excel aided calculations) 'A'=
1	Mercury	2.2097e-9	*62,801,558
2	Venus	3.16917E-9	*62,801,558

3	Earth	1.07793E-7	*62,801,558
4	Mars	3.61512E-8	*62,801,558
5	Jupiter	3.52068Exp-7	*62,801,558
6	Saturn	1.59312Exp-7	*62,801,558
7	Uranus	3.09097EXP-8	*62,801,558
8	Neptune	1.60316Exp-8	*62,801,558
9	Pluto	5.04623Exp-10	*62,801,558

It is significant to note that Table T2 implies to "Occam Razor", or simple Proportionality, as discovered by Albert Einstein and Otto Stern. The Author Steven L. Basic has determined an Excel multiplier "A" = 62,801,558, to be the multiplier within the column number 6.

The magnitude of "A" is gravity related. The unit "Graviton", and the "A" possibly related to graviton, may occur in the gravity force fields in a self-consistent manner.

Table 3 show the Author's Data.

Column 6 represents the names of Planets
Column 7 represents 'g" (column 6 * A+3) in m/second ^2
Column 8 are values of observed surface gravities in m/second ^2 for the planets listed

Table 3

No	6	7	8
			Observed surface gravities [m/second ^2]
		g=(column 6 *A+3) [m/second ^2]	
1	Mercury	3.138752	3.728
2	Venus	3.199028	8.86
3	Earth	9.769568	9.81
4	Mars	5.270351	3.727
5	Jupiter	25.110418	25.928
6	Saturn	13.00504	11.369
7	Uranus	4.943	10.89
8	Neptune	4.0068	11.87
9	Pluto	3.031691	4.218

In the table above, a visual observation and mutual comparison between the two columns 7 and 8, displays the gradients, magnitudes, maximums and the minimums, and confirms that this is a solar system of Albert Einstein and Otto Stern in light of Leonard Euler's and Werner von Heisenbergs radial acceleration of the random Zero Point Energy photons.

Conclusions

In his present study the Author was able to direct some light onto previously unknown facts and concepts.

- Conclusion I: the Zero Point Energy wave posses a linear frequency nature
- Conclusion II: the Zero Point Energy fundamental particle posses a nature of a Circular frequency, radians per second, due to vortex-like simultaneous rotation.
- Conclusion III: following the thought of Nikola Tesla, research scientist and philosopher, any rotation and phenomenon of electricity are related as a cause and consequence. The Zero Point Energy Photon is assigned to be of a fundamental particle nature with an equivalent "mass".
- Conclusion IV: the Zero Point Energy fundamental particle unit given name of ZPE-photon.
- Conclusion V: An electrostatic Direct Current field of high Voltage accelerates rotation of the ZPE-photon that tends to advance along a linear trajectory at the speed of velocity of light (Leonard Euler, Werner von Heisenberg, and the contemporary mathematicians.)

- Conclusion VI: the incorporeal Zero Point Energy waves are present thru-out the entire space and all the temperatures, whose equivalent energy is that of the left-hand side of Louis de Broglie equation; or, their equivalent corporeal "Mass" multiplied by the square of the velocity of light.

- Conclusion VII: The Zero Point Energy waves are immaterial, although the future energy may be based on principles of fuel-less mechanical energy amplifiers.

- Conclusion VIII: The Author's analytical results on the solar system planets and the energy absorbed by the planets, lead him to assertion that the origin and cause of gravitation are parallel to the nature of the Zero Point Energy.

- Conclusion IX: The Zero Point Energy Photons of inter-gravitational field may in fact be related to graviton unit, capable of spin angular momentum throughout the Universe.

- Conclusion X: despite the fact that a cause of gravity is highly complex, its fundamental particles are likely to obey the rules of chromo-dynamics.

- Conclusion XI: the gravitons belong to boson group of fundamental particles. The dynamic nature within the Universe does not exclude possibility that the Zero Point Energy particles are the co-particles of graviton having a spin of 2, a boson rotating about its own axis.

- Conclusion XII: within the space-time in the Universe the Zero Point Energy waves gave rise to itself.

- Conclusion XIII: the intensities of all surface gravities of Celestial Spheres throughout the Solar system are Causal—Consequence of a rotating Zero Point Energy Photon Emission from the Sun.

Gravitational Waves

The main variable of the expression for gravitational waves is the frequency "f", which in the formula stands as the frequency raised to the third degree (or, cube).

An accurate knowledge of this frequency is highly essential.

Consequently, the gravity waves could be counter-illuminated with a similar wave that has one-half wave-length phase displacement, so that the positive amplitude of the gravity wave finds itself across the negative amplitude of the second illuminating wave.

The finding of the exact magnitude of this frequency has not been communicated to other scientist in any monograph, a mathematical analysis will be shown upon a request.

However, the author described gravity (or, Zero Point energy wave) region, this frequency can-not be obtained by a Tesla Coil—Capacitor Resonance.

To Author's liking, considering the principle if anything rotates within the structure of the "ZPE" photon, then, its advancing lobe moves with somewhat greater velocity than the velocity of light.

Simply stated, it can-not be true that the gravitational waves will be controlled before understanding their origin and an adequate mechanical model.

An asserted mechanical model may be that the Zero Point Energy waves are present throughout the entire space and reach nuclei of the planetary matter.

At the nuclear radius level, the nuclei of the matter generate repulsive anti—gravity action which reverses its polarity (the logarithmic term insufficiently researched), and at the atomic radius level becomes gravitational attraction.

The Great Inequality of Jupiter and Saturn

The ratio between the Zero Point Energy of Einstein-Stern, per stereo radian, and the energy of Visible Light Photons, also, per stereo radian, reach a maximum magnitude exactly between the planets of Jupiter and Saturn.

This occurrence may well be the cause of the the inner planets (Jupiter, Mars, Earth, Venus and Mercury) accelerating and approaching the Sun, while the outer planets (Saturn, Uranus, Neptune and Pluto), are decelerating and moving in a direction of the Outer Universe.

Reference 1:

Zero Point Energy Per Stereo Radian and the Distribution of Gravitational Acceleration of Planets Throughout the Solar System, Xlibris, by Steven L. Basic, @ 2013.

References

1.) Steven L. Basic. *Unified Theory, 2009.* ISBN 978-1-4349-9649-7, National Union Catalog OCLC. World Cat-Public View. Pending Library of Congress Control number.

2.) Steven L. Basic. *Indian Journal of Theoretical Physics*, Volume 47, No 1, ISSN 0019-5693, 1999, Institute of Theoretical Physics, Bignan Kutir, 4/1 Mohan Lane, Calcutta, 700 004, India, 'Some remarks concerning the distribution of gravitational acceleration and gravitational mass of planets throughout the solar system' by Steven L. Basic, Principal Engineer.

3.) Nick Hebert. *Quantum Reality, Beyond the New Physics and the Meaning of Reality.* Page 75. ISBN 0-385-23569-0. Anchor Books Double Day New York, London, Toronto, Sydney, Auckland.

About the Author

Steven L. Basic holds Master of Science degree in Aerospace engineering from a reputable European University. In his career he worked as a Principal Stress Engineer for the British and American Aerospace industries.

The author has published four studies in the Indian Journal for Theoretical Physics concerning the distribution of gravitational acceleration and gravitational mass of planets throughout the solar system. In this Volume 2, the Author re-enforces his assertion concerning a linkage between the gravitation and rotation. Furthermore, he describes and shows in substantial detail the algorithm as well as his theory of Zero Point Energy, the Uncertainty Principle, and Copenhagen Interpretation of Quantum Theory within the Outer Universe. This algorithm he also applies to a worldwide well known 'down to Earth', Over Unity Machine of an Academician, as an open, repeatable, mathematical analysis-numerical example. This machine oscillates for two hours without incorporation of the magnetic levitation bearings, to be developed in the future.

The second volume elevates the presented "Occam Razor" toward a scientific assertion. This work is a result of many years of scientific consideration in fundamental physics and mathematical string theory realization.

www.ingramcontent.com/pod-product-compliance
Lightning Source LLC
Chambersburg PA
CBHW022129170526
45157CB00004B/1811